COOL
THANKSGIVING DINNER

Beyond the Basics for Kids Who Cook

LISA
WAGNER

A Division of ABDO

ABDO
Publishing Company

Visit us at www.abdopublishing.com

Published by ABDO Publishing Company, P.O. Box 398166, Minneapolis, MN 55439. Copyright ©2014 by Abdo Consulting Group, Inc. International copyrights reserved in all countries. No part of this book may be reproduced in any form without written permission from the publisher. The Checkerboard Library™ is a trademark and logo of ABDO Publishing Company.

Printed in the United States of America,
North Mankato, Minnesota
102013
012014

 PRINTED ON RECYCLED PAPER

Editor: Liz Salzmann
Content Developer: Nancy Tuminelly
Cover and Interior Design and Production:
Colleen Dolphin, Mighty Media, Inc.
Food Production: Desirée Bussiere
Photo Credits: Colleen Dolphin, Shutterstock

Library of Congress Cataloging-in-Publication Data
Wagner, Lisa, 1958- author.
  Cool Thanksgiving dinner : beyond the basics for kids who cook / Lisa Wagner.
     pages cm. -- (Cool young chefs)
  Includes index.
  Audience: Ages 8 to 12.
  ISBN 978-1-62403-090-1
  1. Thanksgiving cooking--Juvenile literature. 2. Thanksgiving Day--Juvenile literature. I. Title.
  TX739.2.T45W34 2014
  641.5'68--dc23
                          2013022530

TO ADULT HELPERS

Congratulations on being the proud parent of an up-and-coming chef! This series of books is designed for children who have already done some cooking—most likely with your guidance and encouragement. Now, with some of the basics out of the way, it's time to really get cooking!

The focus of this series is on parties and special events. The "Big Idea" is all about the creative side of cooking (mastering a basic method or recipe and then using substitutions to create original recipes). Listening to your young chef's ideas for new creations and sharing your own ideas and experiences can lead to exciting (and delicious) discoveries!

While the recipes are designed to let children cook independently as much as possible, you'll need to set some ground rules for using the kitchen, tools, and ingredients. Most importantly, adult supervision is a must whenever a child uses the stove, oven, or sharp tools. Look for these symbols:

Your assistance, patience, and praise will pay off with tasty rewards for the family, and invaluable life skills for your child. Let the adventures in cooking beyond the basics begin!

CONTENTS

HOST THANKSGIVING DINNER!

Welcome to Cool Young Chefs! If you have already used other Cool Cooking books, this series is for you. You know how to read a recipe and how to prepare ingredients. You have learned about measuring, cooking tools, and kitchen safety. Best of all, you like to cook!

YOUR SKILLS MAKE THE PARTY SPECIAL

Thanksgiving celebrations offer a great chance to use your cooking skills. If the party is at your house, you can use your skills as a host too. Set the dining table ahead of time. Make a **centerpiece** out of gourds and fall flowers. Make place cards for the table. Set up some games to keep the young children entertained. Keep smiling and make everyone feel welcome!

MANY DISHES, MANY COOKS

A traditional Thanksgiving feast includes many different foods. Most of the time, a Thanksgiving meal is a group effort. Everyone makes a special dish to bring to the meal. The best and most popular dishes take on the cook's name. For example, Auntie's Green Beans, or Lulu's Pumpkin Pie, or Papa Jack's Gravy. Make a **delicious** dish and you'll join the club of famous family cooks!

WHAT'S THE BIG IDEA?

Besides being a good cook, a chef is prepared, **efficient**, organized, resourceful, creative, and adventurous. The Big Idea in *Cool Thanksgiving Dinner* is all about being organized.

Almost every Thanksgiving meal includes a roast turkey and many side dishes. It is a real skill to have everything ready at the same time. Being well-organized is especially important.

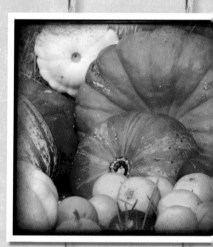

Review the recipes you'll be making for Thanksgiving. Make a list of all the ingredients you'll need. Do most of the shopping several days beforehand. If you're making turkey be sure to give it time to thaw if it is frozen. Some turkeys can take up to a week to thaw. Check the label on the turkey to find out. The label will also tell you how long it takes to cook.

It's a good idea to make whatever you can ahead of time. Things such as cranberry sauce and pies can be made the day before Thanksgiving. You could set the table the day before too.

The best way to stay organized is with a written schedule. Write down when dishes need to go into the oven in order to be ready at dinner time. Write down specific times for starting all of them. Plan everything before you start.

FIRST THINGS FIRST

A successful chef is smart, careful, and patient. Take time to review the basics before you start cooking. After that get creative and have some fun!

BE SMART, BE SAFE

- Start with clean hands, tools, and work surfaces.
- Always get **permission** to use the kitchen, cooking tools, and ingredients.
- Ask an adult when you need help or have questions.
- Always have an adult nearby when you use the stove, oven, or sharp tools.
- Prevent accidents by working slowly and carefully.

NO GERMS ALLOWED

After you handle raw eggs or raw meat, wash your hands with soap and water. Wash tools and work surfaces with soap and water too. Raw eggs and raw meat have bacteria that don't survive when the food is cooked. But the bacteria can survive at room or body temperature. These bacteria can make you very sick if you consume them. So, keep everything clean!

BE PREPARED

- Read through the entire recipe before you do anything else!
- Gather all the tools and ingredients you will need.
- Wash fruits and vegetables well. Pat them dry with a **towel**.
- Get the ingredients ready. The list of ingredients tells how to prepare each item.
- If you see a word you don't know, check the **glossary** on page 30.
- Do the steps in the order they are listed.

GOOD COOKING TAKES TIME

- Allow plenty of time to prepare your recipes.
- Be patient with yourself. **Prep** work can take a long time at first.

ONE LAST THING

- When you are done cooking wash all the dishes and **utensils**.
- Clean up your work area and put away any unused ingredients.

KEY SYMBOLS

In this book, you will see some symbols beside the recipes. Here is what they mean.

The recipe requires the use of a stove or oven. You need adult **supervision** and assistance.

A sharp tool such as a peeler, knife, or **grater** is needed. Be extra careful, and get an adult to stand by.

BEYOND COOL

Remember the Big Idea? In the Beyond Cool boxes, you will find ideas to help you create your own recipes. Once you learn a recipe, you will be able to make many **versions** of it. Remember, being able to make original recipes turns cooks into chefs!

When you modify a recipe, be sure to write down what you did. If anyone asks for your recipe, you will be able to share it proudly.

GET THE PICTURE

When a step number in a recipe has a circle around it with an arrow, it will point to the picture that shows how to do the step.

③ →

COOL TIP

These tips can help you do something faster, better, or more easily.

UNIQUELY COOL

Here are some of the **techniques**, ingredients, and tools used in this book.

TECHNIQUES:

GET FRESH!

Dried herbs are stronger than fresh herbs. If you substitute fresh herbs for dried herbs, use at least three times as much as the recipe calls for. For example, if the recipe says 1 teaspoon dried basil, use 3 teaspoons chopped fresh basil.

MAKE YOUR OWN DRY BREAD CUBES

1 Cut bread into ½-inch cubes.

2 Spread the bread cubes out on baking sheets. Let them dry for a few days.

3 Turn the bread cubes twice a day. The bread will shrink a little as it dries. To get 8 cups of dry bread cubes, dry 10 cups of fresh bread cubes. You might have a little extra, but you won't be short. French bread makes the best bread cubes, but any bread will work.

INGREDIENTS:

GROUND
CINNAMON

GROUND
GINGER

GROUND
NUTMEG

PECANS

TOOLS:

COLANDER

GLASS
MEASURING CUP

GRATER

OVENPROOF
MEAT THERMOMETER

POTATO MASHER

RUBBER SPATULA

SERRATED KNIFE

ZESTER

TURKEY ROASTING TIPS

HOT STUFF! SUPER SHARP!

COOL TIP

Roasting a turkey is a big job. Turkeys are heavy! An adult will probably be in charge of roasting the turkey. With these tips, you can help make sure the turkey is the best one ever.

FRESH TURKEY

- Buy the turkey 1 to 2 days before you plan to cook it.

- Store it in the refrigerator until you're ready to cook it. Place it on a tray to catch any leaking juices.

FROZEN TURKEY

- If the turkey is frozen, put it in the refrigerator to **thaw**. Allow 24 hours for every 4 to 5 pounds of the turkey's weight. It takes a long time to thaw a big bird!

TIMETABLE FOR ROASTING AN UNSTUFFED TURKEY
(AT 325 DEGREES)

8 to 12 pounds	2¾ to 3 hours
12 to 14 pounds	3 to 3¾ hours
14 to 18 pounds	3¾ to 4¼ hours
18 to 20 pounds	4¼ to 4½ hours
20 to 24 pounds	4½ to 5 hours

ROASTING THE TURKEY

- Remove the **giblets** from inside the turkey. Rinse the turkey inside and out and pat dry with paper **towels**. Sprinkle salt and pepper inside the turkey.

- Place the turkey on a rack in a roasting pan. Brush the skin with oil or melted butter. Put 1 cup of water in the roasting pan.

- Stick an ovenproof meat thermometer into the turkey. The thermometer should be placed in the thickest part of the inner thigh. It should point toward the body and should not touch the bone.

- Bake the turkey until the skin is a light golden color. Then place a foil tent over the top of the turkey. During the last 45 minutes of baking, remove the foil tent to brown the skin.

- How do you know when it's ready? Check the thermometer. When it reaches 165°F, the turkey should be done. Test the thickest part of the breast to be sure the thermometer reads 165°F there too.

- Remove the turkey from the oven. Let it sit at room temperature for 20 minutes. This gives the juices time to settle and makes carving easier.

- Remember to always wash anything that comes in contact with raw turkey and its juices with soap and water. This includes your hands, work surfaces, sink, dishes, and **utensils**.

CLASSIC & DELICIOUS DRESSING

ingredients

12 tablespoons butter

1 onion, chopped

4 stalks celery with leaves, chopped

8 cups plain bread cubes

1 tablespoon poultry seasoning

1 cup or more chicken broth

salt and pepper

tools

measuring cups

measuring spoons

sharp knife

cutting board

large pot

wooden spoon

9 × 13-inch baking dish

aluminum foil

1 Preheat the oven to 325 degrees. Grease the inside of the baking dish.

2 Melt the butter in a large pot over medium heat. Add the onion and celery and sauté until softened.

3 Stir in the bread cubes and poultry seasoning. Toss until everything is well blended. Add the broth a little at a time to moisten bread cubes. Add salt and pepper to taste.

4 Put the **dressing** in the baking dish. Cover it with aluminum foil. Bake covered for 20 minutes. Remove the foil and bake for another 20 to 25 minutes.

COOL TIP

Dressing baked inside a turkey is called stuffing. Baking the dressing separately is much easier. You can still bake dressing and roast the turkey at the same time. They use the same oven temperature.

BEYOND COOL

- Poultry seasoning is a blend of sage, thyme, marjoram, and rosemary. Make your own poultry seasoning using these herbs. Use twice as much sage as the other herbs.

- Make your own dry bread cubes. See the instructions on page 10.

- Try making sausage dressing. Use only 7 cups of bread cubes. Sauté 1 pound of breakfast sausage until cooked through. Drain the fat. Add the sausage to the bread cubes before adding the broth.

- Like a dressing with crunch? Add 1 cup of chopped pecans before adding the broth.

CREAMY MASHED POTATOES

ingredients

3 pounds Yukon Gold potatoes, peeled and cut in half

6 tablespoons butter

1½ cups whole milk, warmed

salt and pepper

tools

vegetable peeler

sharp knife

cutting board

measuring cups

large pot

strainer

2 microwave-safe bowls

large mixing bowl

potato masher

mixing spoon

1. Put the potatoes in a large pot. Add cold water until the potatoes are covered. Bring to a boil over high heat. Cook for 10 to 12 minutes until the potatoes are tender but not mushy. Remove from heat and drain potatoes. Do not rinse them.

2. Put the butter in a microwave-safe bowl. Melt it in the microwave using low power for 30 seconds. Put the milk in another microwave-safe bowl and heat for 1 minute using high power.

3. Place the hot potatoes in a large mixing bowl. Mash them until most of the lumps are gone.

4. Stir the butter into the potatoes.

5. Stir in half of the warmed milk. Continue adding milk until the mashed potatoes are smooth and creamy. Add salt and pepper to taste. Top with perfect gravy [see page 18].

BEYOND COOL

For richer mashed potatoes, substitute light or heavy cream for the milk.

COOL TIP

Check the potatoes for doneness by sticking a sharp knife into one. If the knife goes in easily the potatoes are ready.

PLENTY OF PERFECT GRAVY

MAKES 4 CUPS

ingredients

3 tablespoons butter

3 tablespoons flour

4 cups chicken or turkey broth, warmed

poultry seasoning

salt and pepper

tools

measuring spoons

measuring cups

saucepan

whisk

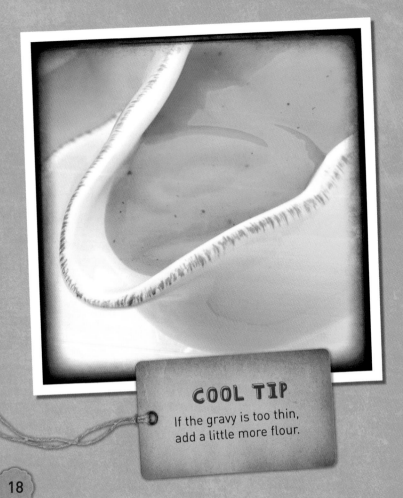

COOL TIP
If the gravy is too thin, add a little more flour.

1 Melt the butter in a saucepan over medium heat. Sprinkle the flour over the butter and whisk to blend. Continue to whisk and cook for 1 minute.

2 Slowly add the broth and whisk until the mixture is smooth.

3 Turn the heat to low. Add poultry seasoning and salt and pepper to taste. Cook for 5 more minutes, stirring from time to time.

ingredients

12 ounces whole cranberries
½ cup white sugar
½ cup brown sugar
zest from 1 orange
1 cup fresh orange juice
pinch of ground cinnamon

tools

measuring cups
zester or grater
saucepan
mixing spoon

1 Combine all the ingredients in a saucepan over medium heat. Stir until the sugar is **dissolved**.

2 Cook and stir often until the cranberries begin to pop. This will take about 10 minutes. If you want a thicker sauce, cook and stir 5 more minutes.

3 Put the sauce in a serving bowl. Refrigerate overnight.

CRANBERRY ORANGE SAUCE

FABULOUS FLORNEY CORN

MAKES
2 CUPS

ingredients

4 eggs

½ cup butter

½ cup maple syrup

2 10-ounce bags of frozen
  corn kernels

tools

measuring cups

9 ×13-inch baking dish

large mixing bowl

hand mixer

spatula

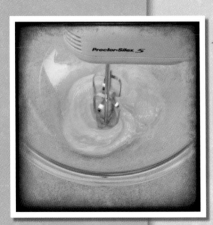

1 Preheat the oven to 350 degrees. Lightly grease the baking dish.

2 Break the eggs into a large mixing bowl. Use a hand mixer to beat the eggs until they are foamy.

3 Melt the butter in the microwave using low power for 30 seconds. Add it to the eggs.

4 Beat in the maple syrup.

5 Spread the frozen corn **kernels** in the baking dish.

6 Pour the egg mixture over the frozen corn. Mix lightly with a spatula.

7 Bake for 1 hour until crusty brown. Stir once halfway through the baking time.

CRUNCHY GLAZED CARROTS

ingredients

4 cups peeled, sliced carrots

⅓ cup fresh orange juice

1 tablespoon cornstarch

2 tablespoons maple syrup

salt and pepper

3 tablespoons chopped fresh parsley

tools

sharp knife

cutting board

measuring cups

measuring spoons

saucepan

strainer

whisk

mixing spoon

serving bowl

1 Put the carrots in a saucepan with 1 inch of water. Bring to a boil. Cover the pan and cook over low heat until the carrots are tender. It takes about 8 minutes. Drain the carrots.

2 Put the orange juice and cornstarch in the saucepan. Whisk until smooth. Add the maple syrup. Bring to a boil. Stir and boil for 1 minute.

3 Return the carrots to the saucepan. Stir and cook over low heat until the carrots are hot. Add salt and pepper to taste.

4 Put the carrots in a serving bowl and sprinkle the chopped parsley on top.

COOL TIP

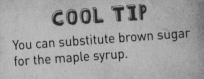

You can substitute brown sugar for the maple syrup.

BEYOND COOL

- For spiced **glazed** carrots, use a pinch of ground nutmeg and ¼ teaspoon of ground ginger. Add the spices with the maple syrup.

- Substitute 1 teaspoon dried or 1 tablespoon fresh thyme for the parsley. If you use dried thyme, add it with the maple syrup.

GREEN BEANS WITH ALMONDS

SERVES 6

ingredients

1 pound fresh green beans, ends trimmed

2 tablespoons butter

2 tablespoons olive oil

½ cup slivered almonds

salt and pepper

tools

sharp knife

cutting board

measuring spoons

measuring cups

large saucepan

strainer

frying pan

rubber spatula

1 Fill a large saucepan halfway with water. Bring it to a boil. Add the beans and cook for 5 minutes. Drain the beans and rinse them with cold water.

2 Heat the butter and olive oil in a frying pan over medium heat. When the butter is melted, add the almonds and turn heat to low. Sauté almonds until they are lightly browned.

3 Put the beans in the frying pan. Sauté for 5 minutes until the beans are tender. Add salt and pepper to taste.

BEYOND COOL

- Cook 4 slices of bacon. Drain on paper **towels**. When the bacon is cooled, chop or **crumble** it into small pieces. Add the bacon just before you add salt and pepper.

- Add chopped fresh herbs such as parsley, dill, or oregano. Toss them in the frying pan with the cooked beans before you add salt and pepper.

COOL TIP

You can use a 16-ounce package of frozen green beans instead of fresh green beans. Then in step 1 cook the beans for 3 minutes instead of 5 minutes.

NUTTY GOODNESS PECAN PIE

ingredients

9-inch uncooked pie crust (purchased or homemade)

½ cup butter

2 eggs, at room temperature

1 cup brown sugar

¼ cup white sugar

1 tablespoon flour

1 tablespoon milk

1 teaspoon vanilla extract

1½ cups chopped pecans

tools

measuring cups

measuring spoons

sharp knife

cutting board

pie plate

fork

large mixing bowl

hand mixer

mixing spoon

1 Preheat the oven to 400 degrees.

2 Place the pie crust in the pie plate. Press lightly to fit it into the bottom of the plate. Trim the edge evenly around the pie plate. Pinch the edge to form a rim around the pie.

3 Use a fork to poke some holes in the bottom of the pie crust.

4 Melt the butter in the microwave.

5 Crack the eggs into a large mixing bowl. Beat them with a hand mixer. When the eggs are foamy, mix in the butter. Add the brown sugar, white sugar, flour, milk, and vanilla. Mix well.

6 Add the pecans. Gently stir them into the mixture.

7 Pour the mixture into the pie crust. Bake for 10 minutes. Turn the oven temperature down to 300 degrees. Bake for 45 to 55 minutes until firm.

COOL TIP
Put a baking sheet on the bottom rack of the oven. It will catch any drips from the pie as it bakes.

SWEET & TART DUTCH APPLE PIE

SERVES 8

ingredients

PIE

9-inch uncooked pie crust (purchased or homemade)

6 cups peeled, cored, sliced Granny Smith apples

1 tablespoon lemon juice

½ cup white sugar

¼ cup brown sugar

2 tablespoons flour

1 teaspoon ground cinnamon

¼ teaspoon ground nutmeg

TOPPING

¾ cup brown sugar

1½ tablespoons white sugar

½ cup flour

4½ tablespoons butter, at room temperature

tools

peeler

sharp knife

cutting board

measuring cups

measuring spoons

pie plate

mixing bowls

mixing spoon

fork

1 Preheat the oven to 375 degrees.

2 Place the pie crust in the pie plate. Press lightly to fit it into the bottom of the plate. Trim the edge evenly around the pie plate. Pinch the edge to form a rim around the pie.

3 Use a fork to poke some holes in the bottom of the pie crust.

4 Put the apples, lemon juice, sugars, flour, cinnamon, and nutmeg in a large mixing bowl. Stir until well blended. Pour the apple mixture into the pie crust.

5 Put all the topping ingredients in a medium mixing bowl. Mash and stir with a fork until evenly mixed.

6 Spread the topping evenly over the apples. Bake for 45 to 55 minutes.

COOL TIP

Put a baking sheet on the bottom rack of the oven. It will catch any drips from the pie as it bakes.

GLOSSARY

centerpiece – a decoration, such as flowers or candles, in the center of a table.

crumble – to break into small pieces.

delicious – very pleasing to taste or smell.

dissolve – to mix with a liquid so that it becomes part of the liquid.

dressing – a seasoned mixture used as a stuffing or baked separately.

efficient – able to do something without wasting time, money, or energy.

giblets – the edible organs of a fowl, such as the heart and liver.

glazed – covered with a thin coating of something.

glossary – a list of the hard or unusual words found in a book.

grater – a tool with sharp-edged holes.

kernel – a grain or seed of a plant such as corn, wheat, or oats.

permission – when a person in charge says it's okay to do something.

prep – short for preparation, the work done before starting to make a recipe, such as washing fruits and vegetables, measuring, cutting, peeling, and grating.

supervision – the act of watching over or directing others.

technique – a method or style in which something is done.

thaw – to melt or unfreeze.

WEB SITES

towel – a cloth or paper used for cleaning or drying.

utensil – a tool used to prepare or eat food.

version – a different form or type from the original.

To learn more about cool cooking, visit ABDO Publishing Company online at www.abdopublishing.com. Web sites about cool cooking are featured on our Book Links page. These links are monitored and updated to provide the most current information available.

INDEX